AF372995

THE FUTURE IS STEM:
AN HBCU ALPHABET

LESLIE D. W. JONES

ISBN: 979-8-9953805-2-8
Front cover image by Leslie D. W. Jones.
First printing edition 2026.
Published by The Hundred-Seven
contact@thehundred-seven.org

Dedicated to the brilliant STEM leaders whose discoveries, courage, and vision continue to inspire the world. Your excellence lights the path for future innovators.

And to the children who read these pages—may your curiosity grow, your imagination soar, and your dreams lead you to create, explore, and discover. The future of STEM is in your hands.

A is for Astronauts

An **astronaut** is a scientist who is trained to travel into space to study planets, stars, and other things in our solar system.

Ronald McNair, an alumnus of **North Carolina A&T State University**, was the second Black person to travel into space.
Christa McAuliffe was the first teacher who was trained to be an astronaut. She earned a master's degree from **Bowie State University**.

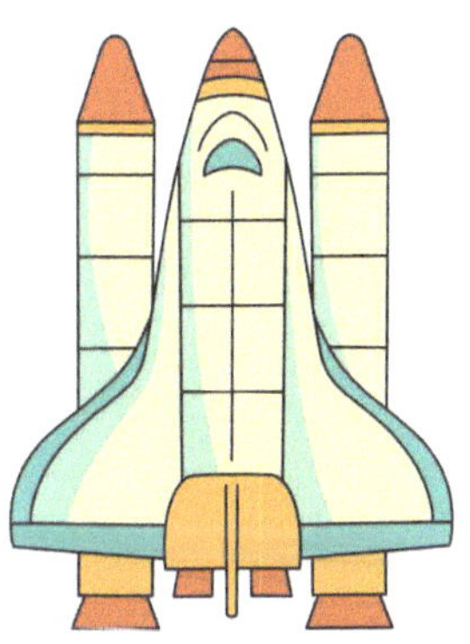

B is for Botany

Botany is the scientific study of plants. A botanist learns about how plants are built, how they reproduce, and how plants interact with the environment.

O'Neil Collins was a **botanist** and professor who graduated from **Southern University**. He was also a mycologist, someone who studies fungi, and a world-renowned expert in the genetics of slime molds.

C is for Chemistry

Chemistry is the science of how things are made and how they react with each other. A **chemist** is a scientist who conducts experiments to learn how different materials work, especially when they are mixed together.

St. Elmo Brady graduated from **Fisk University**. He later became the first African-American to earn a Ph.D. in Chemistry. Brady taught college at several HBCUs and encouraged many students to become scientists.

D is for Discovery

Discovery is an important part of science. Many scientist focus on discovering new things or new ways that matter interacts.

Clarice Phelps is an alumna of **Tennessee State University**. In 2010 she became the first Black woman to help discover a new element-tennessine (element 117)-- the second-heaviest known element. In 1969 and 1970 **Huston-Tillotson University** alumnus **James Harris** was part of a group that discovered elements 104 (rutherfordium) and 105 (dubnium). He is the first Black person to be involved in the discovery of an element.

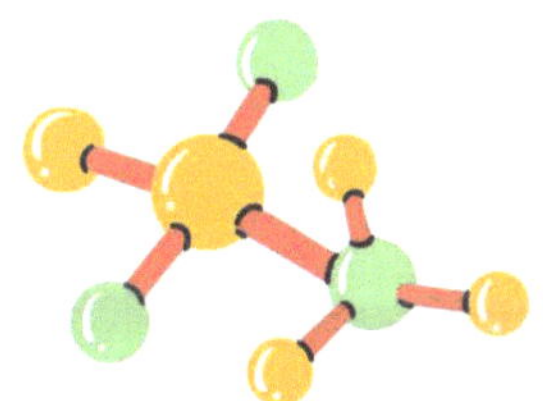

E is for Euphemia Lofton Haynes

Euphemia Lofton Haynes was a **mathematician** and educator. She was the first Black woman to earn a Ph.D. in mathematics.

She taught for 47 years in Washington, DC. In 1966 she became the first woman chairperson of the DC Board of Education. Before becoming a teacher she graduated from what is now known as the **University of the District of Columbia.**

F is for Fossils

Fossils are plants or the hard parts of animals (like bones, shells, or teeth) that turned into stone over thousands of years. They are clues to what life was like millions of years ago.

Paleontologists are scientists who use fossils and other remains to learn about the history of life on Earth.

Aaron Woodard is a paleontologist who graduated from **Fort Valley State University**. **Howard University** alumnus **Louis R. Purnell** was a pioneering paleontologist who specialized in the fossils of invertebrates (animals with no backbone). He was the first Black curator at the Smithsonian.

G is for GPS and Gladys West

GPS (Global Positioning System) is a technology that uses satellites in space to help locate a phone or computer. It is important for helping cell phones work and for navigation systems in cars.

Gladys West was a **mathematician** whose computer programming and precise calculations from satellites were used to create a detailed map of the Earth's shape. Her models helped create the GPS now in everyday use.

H is for Hidden Figures

Hidden Figures are the many Black women who contributed to advances in STEM but were not recognized for their work for many years.

Three of the most well-known Hidden Figures were **Katherine Johnson**, **Mary Jackson**, and **Dorothy Vaughan**. They all worked as human computers at what is now NASA and helped the United States dominate the 'Space Race.' Johnson graduated from **West Virginia State University**, Jackson graduated from **Hampton University** and Vaughan graduated from **Wilberforce University**.

I is for Innovation

Innovation is the result of scientific research. Scientists create something new that makes things easier for people. The **inventor** gets a **patent** saying they own the new creation. HBCUs have researchers who have created new things. More than 30 HBCUs have patents for different innovations. Many HBCU alumni are innovators as well like **Tuskegee University** alumnus **Lonnie Johnson** who has more than 100 patents for his inventions, including the Super Soaker. Businessman **Alfred L. Cralle** attended what is now **Virginia Union University** and invented an ice cream scooper; **Alice H. Parker,** who attended **Howard University**'s high school academy, received a patent for her heating furnace which used natural gas instead of a wood-burning system to send heat through ducts.

J is for Jewel Plummer Cobb

Jewel Plummer Cobb was a researcher and college president. After graduating from **Talladega College**, she studied skin cells that produce **melanin** (what gives skin its color and protects it) and how those cells become cancerous. Her research on **melanoma**, a type of skin cancer, helped scientists create new and more effective treatments.

K is for Samuel Kountz

Samuel Kountz was a trailblazing surgeon from Arkansas. Kountz graduated from the **University of Arkansas at Pine Bluff** and became an expert on **kidneys**, which the body uses to remove waste and extra water from the blood. Kountz performed the first successful kidney transplant between humans who were not identical twins. He also helped create a device that can store kidneys for up to 50 hours to be used for a transplant surgery.

L is for Laboratory

A **laboratory** (lab) is a special room where scientists and students do experiments to learn new things. It has tools like microscopes, test tubes, and computers to help study science. College students use laboratories to practice what they learn in their classes. In the lab they test ideas, observe results, and discover how the world works. HBCUs have labs for all kinds of research including chemistry, cybersecurity, robotics, nursing, and even video games.

M is for Marine Science

Marine science focuses on oceans and the animals and plants that live there. This includes sea animals like fish, whales, dolphins, and turtles. They also study coral reefs, plants in the ocean, and the water itself. By learning about the ocean, they help protect sea life and keep our oceans healthy.

Samuel Nabrit graduated from **Morehouse College**. He became the first African-American to earn a doctoral degree from Brown University and was later a college president. A marine biologist, Nabrit studied how some injured fish are able to regrow their tail fins through a process called regeneration.

N is for NASA

NASA (National Aeronautics and Space Administration) is a government agency that is responsible for carrying out space exploration and research.

Julian Earls, an alumnus of **Norfolk State University**, led NASA's Glenn Research Center in Cleveland, Ohio. He was the first Black man to be appointed section head, office chief, division chief and deputy director of NASA.

O is for Ophthalmology

Ophthalmology is a branch of medicine that studies and treats diseases of the eye.

Charles V. Roman, although raised in Canada, was educated at **Fisk University** and **Meharry Medical College** in Tennessee. He was the first African-American physician to receive training in both ophthalmology and **otolaryngology** (the study of the ear, nose, and throat).

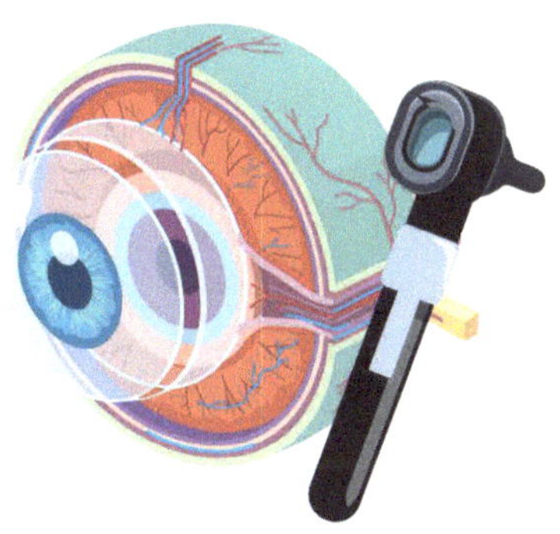

P is for Planetarium

A **planetarium** is a special theater that projects the night sky onto its domed ceiling. It is designed to teach astronomy and space science to people of all ages.

Several HBCUs have a planetarium on their campus including **Fayetteville State University**, **South Carolina State University** and **Elizabeth City State University**. They host special shows for the public throughout the year.

Q is for Lloyd Quarterman

Lloyd Quarterman was a **chemist** whose work focused on gases and how they interact, especially **fluorine**. After graduating from **Saint Augustine's University**, he became one of six Black scientists who helped create the atomic bomb. He later created a new compound made with **xenon**, a gas that scientists once thought could not join with other atoms to form compounds.

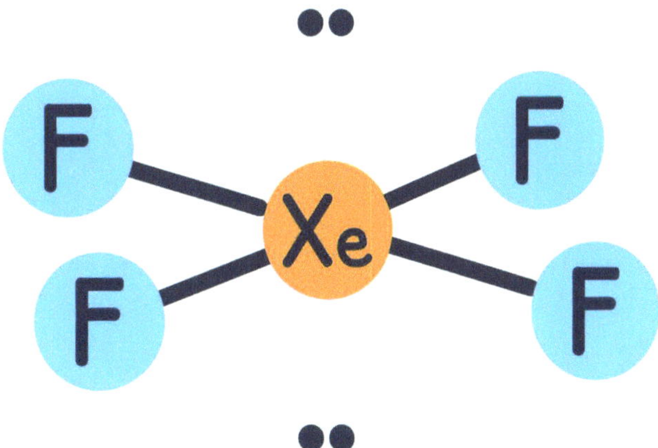

R is for Robotics

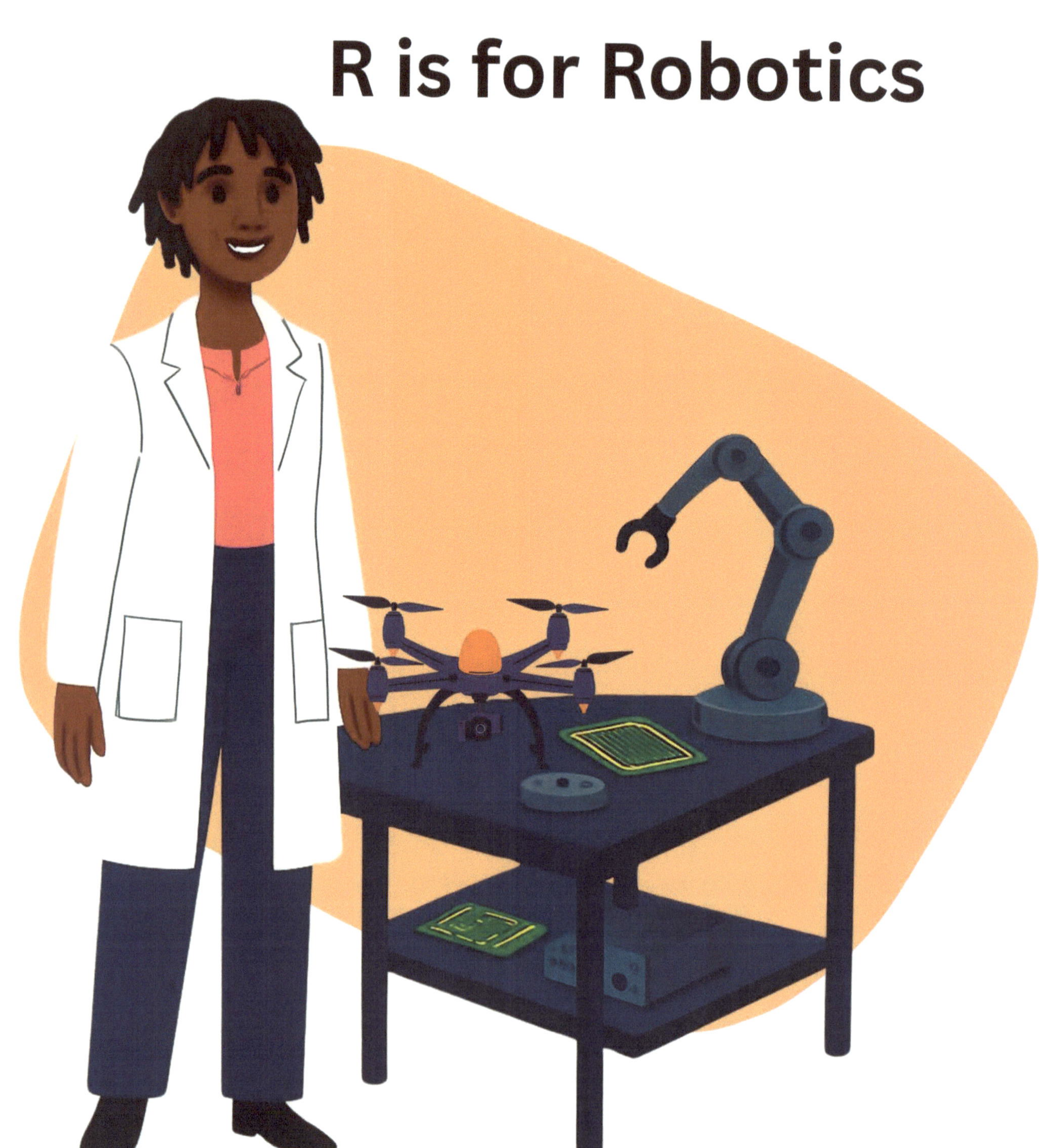

Robotics is a STEM field that combines engineering and computer science to design, build, and program machines to complete complex jobs.

HBCUs offer programs, including a special field called mechatronics, to train students to do that work. Some schools also have robotics teams that design robots for competitions.

Nialah Wilson-Small graduated from **Howard University** and now develops **algorithms** (detailed instructions) for robot systems and human-drone interactions.

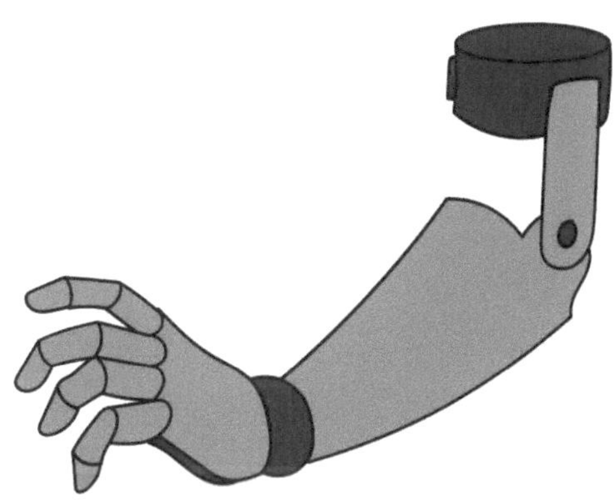

S is for Sunspot

A **sunspot** is an area that appears dark on the surface of the Sun. They appear dark because they are cooler than other parts of the Sun's surface.

After earning a degree in mathematics from **Alabama A&M University**, **Jeanette Scissum** became the first Black woman mathematician and scientist at NASA Marshall Space Flight Center. Her research improved techniques for forecasting the sunspot cycle.

T is for Telescope

A **telescope** is a tool that uses lenses and mirrors to help make distant objects like stars and planets appear brighter, clearer, and closer.

The James Webb Space Telescope is the largest telescope in space. It is able to send images from distant galaxies back to Earth as it travels. **Gregory Robinson**, an alumnus of **Virginia Union University** and **Howard University** was the director of the Webb program from 2018 through its launch.

Astrophysicist **Hakeem Oluseyi** graduated from **Tougaloo College**. He helped design, build, calibrate, and launch the Multi-Spectral Solar Telescope Array which was launched on a rocket and took images of the Sun.

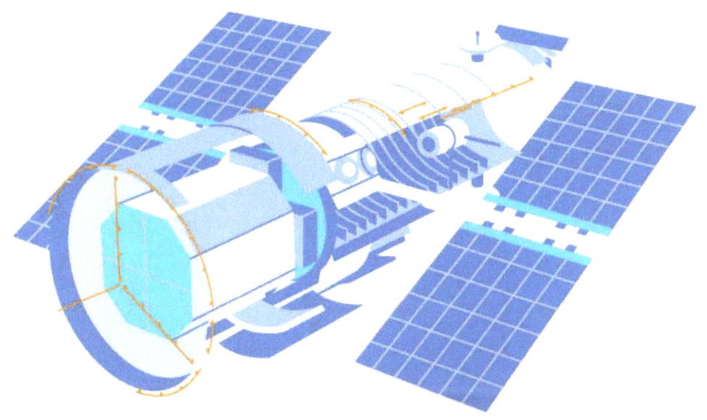

U is for Universe

The **universe** is absolutely everything—all the matter, energy, stars, planets, and galaxies, plus all time and space.

Students interested in studying the universe can consider astronomy and planetary science degree programs offered at **Hampton University**, **Howard University**, **South Carolina State University**, and the **University of the Virgin Islands**.

V is for Veterinarian

Veterinarians are doctors for animals. They study veterinary science in school and are able to treat all kinds of animals from house pets to animals on farms and in zoos. **Tuskegee University** established the first and only HBCU veterinary school in 1945. Seventy-five percent of all Black veterinarians graduated from Tuskegee. The **University of Maryland Eastern Shore** is establishing a new School of Veterinary Medicine and will be the second HBCU with the program.

W is for the Wizard of Tuskegee

Botanist and agricultural teacher **George Washington Carver** was called the **Wizard of Tuskegee** due to his pioneering research with plants including soybeans, sweet potatoes, and peanuts. Carver saved the agriculture of the American South through the development of a crop rotation method. Visitors to the **Tuskegee University** campus can visit his laboratory.

X is for X-ray

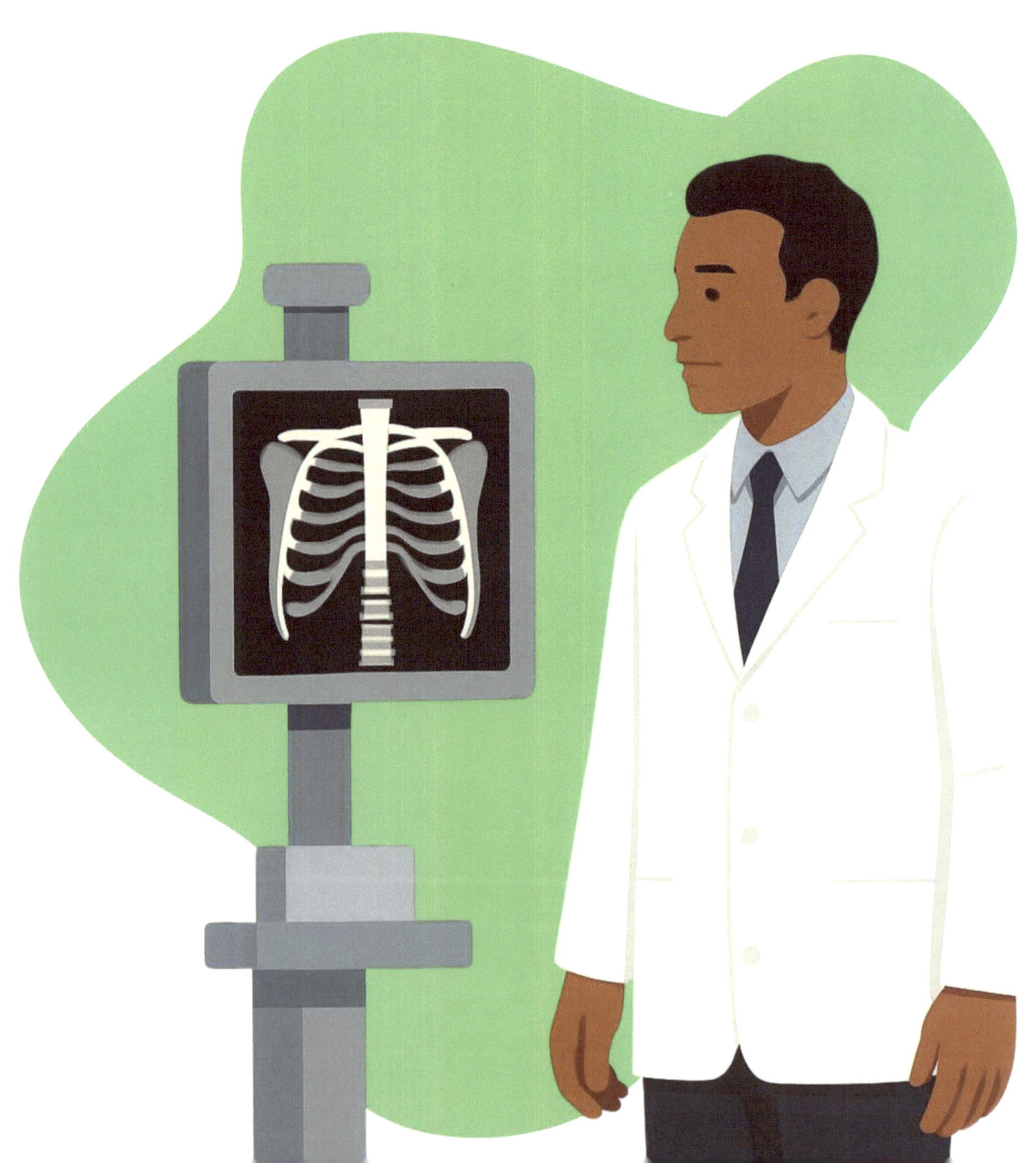

An **X-ray** is an imaging test that shows pictures of the bones and soft tissues inside of the body. X-rays are used to detect breaks in bones as well as illnesses and tumors (abnormal tissue growth).

C.B. Powell was a **Howard University**-trained physician who operated an X-ray lab in Harlem, New York. He was one of the first Black doctors to specialize in X-rays.

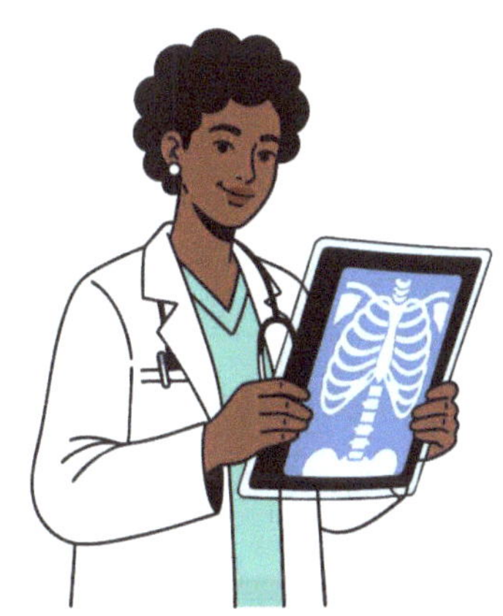

Y is for Roger Arliner Young

Roger Arliner Young was the first Black woman to earn a doctorate degree in **zoology**. Young was also the first Black woman admitted to Sigma Xi scientific research society, and the first to publish original research in her field. She was a professor at several HBCUs including **North Carolina Central University**, **Jackson State University**, and her alma mater, **Howard University**.

Z is for Zoology

Zoology is a branch of biology that studies animals.

Ernest E. Just was a **zoologist** whose research focused on marine animals. He attended high school at what is now **South Carolina State University** and eventually became head of the Zoology Department at **Howard University**.

STEM Vocabulary

algorithm	melanin
astronaut	melanoma
botanist	NASA
botany	ophthalmology
chemist	otolaryngology
chemistry	paleontologist
discovery	patent
fluorine	planetarium
fossil	robotics
GPS	sunspot
innovation	telescope
inventor	universe
kidneys	veterinarian
laboratory	X-ray
marine science	xenon
mathematician	zoology

HBCU STEM Legends

featured in this book

Brady, St.Elmo
Carver, George Washington
Cobb, Jewel Plummer
Collins, O'Neil
Cralle, Alfred
Earls, Julian
Harris, James
Haynes, Euphemia Lofton
Just, Ernest E.
Johnson, Katherine
Kountz, Samuel
Vaughan, Dorothy
Jackson, Mary
Johnson, Lonnie

McAuliffe, Christa
McNair, Ronald
Nabrit, Samuel
Parker, Alice H.
Phelps, Clarice
Powell, C.B.
Purnell, Louis R.
Scissum, Jeanette
Quarterman, Lloyd
West, Gladys
Wilson-Small, Nialah
Woodard, Aaron
Young, Roger Arliner